ENERGY SECTOR STANDARD
OF THE PEOPLE' S REPUBLIC OF CHINA

中华人民共和国能源行业标准

Occupational Safety and Health Acceptance Specification for Wind Power Projects

风电场工程劳动安全与工业卫生验收规程

NB/T 31073-2015

Chief Development Organization: China Renewable Energy Engineering Institute

Approval Department: National Energy Administration of the People's Republic of China

Implementation Date: September 1, 2015

China Water & Power Press

中国水利水电出版社

Beijing 2024

图书在版编目（CIP）数据
风电场工程劳动安全与工业卫生验收规程 : NB/T 31073-2015 = Occupational Safety and Health Acceptance Specification for Wind Power Projects (NB/T 31073-2015) : 英文 / 国家能源局发布. -- 北京 : 中国水利水电出版社, 2024. 10. -- ISBN 978-7-5226-2772-4
Ⅰ. TM614-65
中国国家版本馆CIP数据核字第2024P3C009号

ENERGY SECTOR STANDARD
OF THE PEOPLE'S REPUBLIC OF CHINA
中华人民共和国能源行业标准

Occupational Safety and Health Acceptance Specification for Wind Power Projects

风电场工程劳动安全与工业卫生验收规程

NB/T 31073-2015

（英文版）

Issued by National Energy Administration of the People's Republic of China
国家能源局　发布
Translation organized by China Renewable Energy Engineering Institute
水电水利规划设计总院　组织翻译
Published by China Water & Power Press
中国水利水电出版社　出版发行
Tel: (+ 86 10) 68545888　68545874
sales@mwr.gov.cn
Account name: China Water & Power Press
Address: No.1, Yuyuantan Nanlu, Haidian District, Beijing 100038, China
http: //www.waterpub.com.cn
中国水利水电出版社微机排版中心　排版
北京中献拓方科技发展有限公司　印刷
184mm×260mm　16 开本　2.75 印张　87 千字
2024 年 10 月第 1 版　2024 年 10 月第 1 次印刷
Price（定价）: ¥ 440.00

Introduction

This English version is one of China's energy sector standard series in English. Its translation was organized by China Renewable Energy Engineering Institute authorized by National Energy Administration of the People's Republic of China in compliance with relevant procedures and stipulations. This English version was issued by National Energy Administration of the People's Republic of China in Announcement [2023] No. 5, dated October 11, 2023.

This version was translated from the Chinese Standard NB/T 31073-2015, *Occupational Safety and Health Acceptance Specification for Wind Power Projects*, published by China Water & Power Press. The copyright is reserved by National Energy Administration of the People's Republic of China. In the event of any discrepancy in the implementation, the Chinese version shall prevail.

Many thanks go to the staff from the relevant standard development organizations and those who have provided generous assistance in the translation and review process.

For further improvement of the English version, any comments and suggestions are welcome and should be addressed to:

China Renewable Energy Engineering Institute
No. 2 Beixiaojie, Liupukang, Xicheng District, Beijing 100120, China
Website: www.creei.cn

Translating organizations:

China Renewable Energy Engineering Institute

China Water Resources & Hydropower Construction Engineering Consulting Co., Ltd.

Translating staff:

WANG Jilin JIA Chao YUE Lei

Review panel members:

GUO Jie	POWERCHINA Beijing Engineering Corporation Limited
QIE Chunsheng	Senior English Translator
YAN Wenjun	Army Academy of Armored Forces, PLA
CHEN Lei	POWERCHINA Zhongnan Engineering Corporation Limited

LI Zhongjie	POWERCHINA Northwest Engineering Corporation Limited
CHE Zhenying	IBF Technologies Co., Ltd.

National Energy Administration of the People's Republic of China

翻译出版说明

本译本为国家能源局委托水电水利规划设计总院按照有关程序和规定，统一组织翻译的能源行业标准英文版系列译本之一。2023 年 10 月 11 日，国家能源局以 2023 年第 5 号公告予以公布。

本译本是根据中国电力出版社出版的《风电场工程劳动安全与工业卫生验收规程》NB/T 31073—2015 翻译的，著作权归国家能源局所有。在使用过程中，如出现异议，以中文版为准。

本译本在翻译和审核过程中，本标准编制单位及编制组有关成员给予了积极协助。

为不断提高本译本的质量，欢迎使用者提出意见和建议，并反馈给水电水利规划设计总院。

地址：北京市西城区六铺炕北小街 2 号
邮编：100120
网址：www.creei.cn

本译本翻译单位：水电水利规划设计总院
中国水利水电建设工程咨询有限公司

本译本翻译人员：王继琳　贾　超　岳　蕾

本译本审核成员：

郭　洁　中国电建集团北京勘测设计研究院有限公司
郄春生　英语高级翻译
闫文军　中国人民解放军陆军装甲兵学院
陈　蕾　中国电建集团中南勘测设计研究院有限公司
李仲杰　中国电建集团西北勘测设计研究院有限公司
车振英　一百分信息技术有限公司

国家能源局

Announcement of National Energy Administration of the People's Republic of China [2015] No. 3

According to the requirements of Document GNJKJ [2009] No. 52, "Notice on Releasing the Energy Sector Standardization Administration Regulations (*tentative*) and detailed implementation rules issued by National Energy Administration of the People's Republic of China", 203 sector standards such as *Carbon Steel and Low Alloy Steel for Pressurized Water Reactor Nuclear Power Plants—Part 31: 15Mn Forgings for Containment Vessel*, including 106 energy standards (NB) and 97 electric power standards (DL), are issued by National Energy Administration of the People's Republic of China after due review and approval.

Attachment: Directory of Sector Standards

National Energy Administration of the People's Republic of China

April 2, 2015

Attachment:

Directory of Sector Standards

Serial number	Standard No.	Title	Replaced standard No.	Adopted international standard No.	Approval date	Implementation date
…						
57	NB/T 31073-2015	Occupational Safety and Health Acceptance Specification for Wind Power Projects			2015-04-02	2015-09-01
…						

Foreword

According to the requirements of Document GNKJ [2012] No. 83 issued by National Energy Administration of the People's Republic of China, "Notice on Releasing the Development and Revision Plan of the First Batch of Energy Sector Standards in 2012", and after extensive investigation and research, summarization of practical experience in the occupational safety and health acceptance of wind power projects in recent years, and wide solicitation of opinions, the drafting group has prepared this specification.

The main technical contents of this specification include acceptance conditions, acceptance procedures, main content of acceptance, and preparation requirements and content of acceptance documents.

National Energy Administration of the People's Republic of China is in charge of the administration of this specification. China Renewable Energy Engineering Institute has proposed this specification and is responsible for its routine management. Sub-Committee on Planning and Design of Wind Power Project of Energy Sector Standardization Technical Committee on Wind Power is responsible for the explanation of specific technical contents. Comments and suggestions in the implementation of this specification should be addressed to:

China Renewable Energy Engineering Institute
No. 2 Beixiaojie, Liupukang, Xicheng District, Beijing 100120, China

Chief development organizations:

China Renewable Energy Engineering Institute

Participating development organizations:

POWERCHINA Northwest Engineering Corporation Limited

POWERCHINA Zhongnan Engineering Corporation Limited

Hubei Anyuan Safety & Environmental Protection Technology Co., Ltd.

Chief drafting staff:

NIU Wenbin	CHENG Feng	WANG Jilin	TIAN Zaiwang
LI Hong	LI Jing	WANG Yibing	YAO Yunlong
DAI Xiangrong	ZHAO Xinchang	ZHANG Xiaoli	PAN Jian
ZHANG Xiaoguang	ZENG Hui	JIA Chao	

Review panel members:

XIE Hongwen	YANG Zhigang	FAN Xiaoping	TIAN Dongsheng

ZHENG Xiwen	YAO Shuanxi	LU Zhaoqin	CHEN Yinqi
ZHONG Tao	CHEN Xuan	YANG Jinghui	ZHAO Zhuo
WU Min	LI Angui	ZHAO Zhaofeng	WANG Yi

Contents

Contents

1 General Provisions

1.0.1 This specification is formulated with a view to standardizing and guiding the occupational safety and health acceptance of wind power projects in accordance with the requirements of national laws, regulations, rules, standards, etc.

1.0.2 This specification applies to the occupational safety and health acceptance of the construction, renovation and extension of grid-connected onshore and offshore wind power projects.

1.0.3 The scope of occupational safety and health acceptance of wind power projects includes the safety protection facilities related to the systems such as wind turbines and box-type transformers, collection lines, step-up substations, and other auxiliary facilities of the wind power projects.

1.0.4 The occupational safety and health acceptance of wind power projects is divided into three stages: application for acceptance and data prereview, on-site inspection and joint review, and acceptance.

1.0.5 The occupational safety and health acceptance of wind power projects is carried out by the acceptance committee composed of the acceptance presiding organization, local safety supervision authority, etc.

1.0.6 The project owner shall report to the acceptance presiding organization the use of safety facilities and major incidents and accidents that happen within 2 years after occupational safety and health acceptance of the project.

1.0.7 In addition to this specification, the occupational safety and health acceptance of wind power projects shall comply with other current relevant standards of China.

2 Terms

2.0.1 occupational safety and health

work field aimed to safeguard the safety and health of workers in workplaces and corresponding measures taken in terms of law, technology, equipment, organizational system, education, etc. Synonyms: occupational health and safety, labor safety and health, labor safety and occupational health, and labor protection

2.0.2 project safety facilities

general term for equipment, facilities, installations, buildings (structures) and other technical measures used by the wind power project operator in the production and operation activities to prevent work safety-related accidents

2.0.3 regime of three concurrences

practice that project safety facilities must be designed, constructed and put into operation simultaneously with the main works

3 Acceptance Conditions

3.0.1 The civil works within the scope of acceptance have been completed and put into operation in accordance with the approved documents, the main works is completed and has passed the fire protection acceptance, and all electromechanical equipment has been put into operation for two full months.

3.0.2 If the project is constructed by stages, the works of each stage is completed and has passed the fire protection acceptance.

3.0.3 A safety assessment organization with appropriate qualifications has completed the safety acceptance assessment report and given a clear conclusion that the acceptance conditions have been met.

3.0.4 The acceptance documents and data are complete.

4 Acceptance Procedures

4.1 Acceptance Organization

4.1.1 The acceptance presiding organization is responsible for organizing the acceptance work of occupational safety and health. The acceptance presiding organization, together with the local safety supervision authority and other organizations, forms an acceptance committee and an acceptance experts team. The acceptance committee is responsible for reviewing the acceptance appraisal certificates, and the acceptance experts team performs on-site inspection.

4.1.2 The acceptance committee is composed of the chairman, deputy chairmen and members. The chairman is usually held by a representative of the acceptance presiding organization. The deputy chairmen are held by the representatives from the local safety supervision authority and the project owner's superior authorities. The committee members include the representatives of the construction contractors and operator and the leader of the acceptance experts team.

4.2 Application for Acceptance

4.2.1 When the project meets the conditions for the occupational safety and health acceptance, the project owner shall submit an application for the occupational safety and health acceptance to the acceptance presiding organization according to the acceptance procedures (see Appendix A), and submit the following paper documents and data, and an electronic copy (CD-ROM):

1 The acceptance application report and application form (3 copies), see Appendix B for the application form of acceptance.

2 The design self-inspection report, supervision self-inspection report, construction and installation self-inspection report and operation self-inspection report about the occupational safety and health acceptance (3 copies each).

3 The safety acceptance assessment report (draft) prepared by a qualified safety assessment organization (10 copies).

4.2.2 The project owner shall make the following documents available for reference (the originals if necessary):

1 The safety pre-assessment report and its review opinions.

2 The feasibility study report and its review opinions.

3 The fire protection (or staged project fire protection) acceptance conformity conclusion.

4 The investment in occupational safety and health and its use.

5 The reports on work safety accidents and other major quality-related accidents during construction.

6 The review opinions on major changes in the main works design or safety facilities design involving the engineering safety.

7 Other approval documents, stage acceptance reports, contract documents and drawings, technical design documents, analysis report on safety monitoring results, etc. related to the occupational safety and health acceptance.

8 The mandatory testing and inspection reports, including plant-wide earthing resistance test report, special-purpose equipment inspection and testing reports, testing report of hazardous factors in workplaces, inspection and testing reports of pressure gauges and relief valves, etc.

9 The confirmation documents for rectification of problems presented in the safety acceptance assessment report.

10 The safety management and emergency preparedness plan.

4.3 Data Prereview

4.3.1 The acceptance presiding organization organizes subject matter experts to complete the prereview within 10 working days after receiving the application for occupational safety and health acceptance.

4.3.2 For those that have passed the prereview, the acceptance presiding organization organizes the occupational safety and health acceptance; for those that fail the prereview, the acceptance presiding organization issues a written notification timely, requiring the project owner to supplement and improve the acceptance documentation, and re-organize the prereview.

4.4 On-Site Inspection and Joint Review

4.4.1 Preparations:

1 Before on-site inspection, the acceptance presiding organization shall consult the local safety supervision authority to determine the member organizations of the acceptance committee, and organize subject matter experts to establish an acceptance experts team.

2 The acceptance presiding organization shall prepare an acceptance program before the on-site inspection, and send a copy of the acceptance program to each member organization of acceptance committee. The main content of the acceptance program is shown in Appendix C of this specification.

4.4.2 On-site inspection and joint review:

1 The acceptance experts team conducts the on-site inspection and the technical review of the safety acceptance assessment report (for review), and puts forward the on-site inspection opinions and prereview opinions of safety acceptance assessment report.

2 After the occupational safety and health facilities of wind power projects pass the on-site inspection by acceptance experts team, the project owner shall make rectifications according to the on-site acceptance inspection opinions, and submit a written rectification report to the acceptance presiding organization. When necessary, the acceptance presiding organization shall check the field rectification made by the project owner.

If the occupational safety and health facilities of the wind power project fail to pass the on-site inspection and acceptance, the project owner shall make rectifications in accordance with the relevant national regulations and the on-site inspection opinions, and reassess the safety acceptance before applying for the occupational safety and health acceptance according to the procedures.

3 If the safety acceptance assessment report (for review) passes the technical review, the assessment organization shall revise and improve the report according to the preliminary opinions, and submit the report to the acceptance presiding organization for review.

If the safety acceptance assessment report (for review) fails to pass the technical review, the assessment organization shall conduct the reassessment, and submit it to the acceptance presiding organization for review.

4.5 Review and Acceptance

4.5.1 The acceptance presiding organization shall timely organize the acceptance committee member organizations to hold a meeting on occupational safety and health acceptance after confirming that the project owner has completed the required rectification, and the assessment organization has modified the safety acceptance assessment report (for filing) according to the prereview opinions on the safety acceptance assessment report.

4.5.2 The project owner shall provide the following information for reference at the acceptance meeting:

1 The on-site inspection and rectification report on occupational safety and health acceptance.

2 The safety acceptance assessment report (for filing).

3 The design self-inspection report, supervision self-inspection report, construction and installation self-inspection report and operation self-inspection report about the occupational safety and health acceptance which have been modified according to the opinions of on-site inspection.

4.5.3 The acceptance committee formulates after discussion the appraisal certificate on the occupational safety and health acceptance of wind power projects, see Appendix D of this specification for the certificate format, and the acceptance appraisal certificate shall give clear conclusions. Acceptance conclusions must obtain the consent of at least two-thirds of acceptance committee members. Acceptance committee members shall put their signatures on the appraisal certificate. The members having discrepancy with the acceptance conclusions shall present their opinions on the acceptance appraisal certificate and sign.

4.5.4 The occupational safety and health acceptance is deemed unqualified in any of the following cases:

1 The construction contractor does not have the appropriate qualification for construction.

2 The project construction is not in line with the design documents of engineering safety facilities, or the construction quality does not meet the requirements stated in the design documents of engineering safety facilities.

3 The construction of engineering safety facilities does not meet the requirements of relevant national standards for construction technology.

4 The safety assessment organization does not have the appropriate qualifications for safety acceptance assessment, or the safety acceptance assessment is unqualified.

5 The safety facilities and work safety conditions do not meet the relevant laws, regulations, rules and national standards or sector standards and technical specifications for work safety.

6 There is accident potential found during the trial run of the construction project that have not been rectified.

7 There is no work safety management organization established, or work safety management personnel designated in accordance with the laws.

8 The employees have not received safety education and training or do not have corresponding qualifications.

9 Other provisions of laws and administrative regulations have not been met.

4.5.5 The acceptance presiding organization officially issues the review opinions on safety acceptance assessment report to the project owner, according to the completion acceptance, and copies them to the relevant authorities and organizations.

4.5.6 The appraisal certificate of occupational safety and health acceptance of wind power project is approved by the acceptance presiding organization, and then submitted to the local safety supervision authority and copied to all member organizations of acceptance committee by the acceptance presiding organization.

5 Main Content of Acceptance

5.0.1 The on-site inspection of occupational safety and health facilities mainly includes the following:

1 Work safety conditions and provision and use of safety facilities.

2 General protection facilities and measures for wind power projects, including the geological disaster prevention and control measures, flood (tide) and water-logging prevention measures, meteorological disaster prevention measures, traffic safety facilities for access roads and on-site accesses, etc.

3 Fire prevention and explosion-proof measures.

4 Lightning and electrical injury prevention measures.

5 Measures for preventing mechanical injury, lifting injury and fall from elevated positions.

6 Safety control measures for hazardous factors in workplaces, including noise control measures, anti-vibration measures, dust-proof measures, anti-pollution measures and anti-corrosion measures, anti-electromagnetic radiation measures, daylighting and lighting measures, control measures for indoor air quality, temperature, humidity and toxic and hazardous substances, provision of personal protective equipment, etc.

7 Safety monitoring and testing facilities.

8 Inspection and testing of special-purpose equipment and mandatory inspection equipment and facilities.

9 Safety colors and safety signs.

10 Provision of auxiliary rooms.

11 Establishment of work safety management organization or safety management staffing.

12 Development and implementation of safety management system and emergency preparedness plan.

13 Safety education and training of workers, and qualifications of special operation personnel.

14 Anti-terror facilities and emergency preparedness measures.

15 Investment in occupational safety and health and its use.

5.0.2 The technical review of project safety acceptance assessment report mainly includes the following:

1 Validity of the relevant laws, regulations, and standards cited by the safety acceptance assessment report.

2 Definition of assessment units and their assessment methods.

3 Identification of the main dangerous and hazardous factors and the corresponding compliance assessment.

4 Implementation and rectification of safety measures and recommendations.

5 Whether the assessment conclusions are objective and credible.

6 Preparation Requirements and Content of Acceptance Documents

6.0.1 The design self-inspection report, supervision self-inspection report, construction and installation self-inspection report and operation self-inspection report about the occupational safety and health acceptance are prepared by the designer, supervisior, construction contractor, and the operator, respectively. Each organization shall be responsible for the authenticity of the materials it contains. The formats for acceptance self-inspection report are shown in Appendix E of this specification.

6.0.2 The preparation requirements for and the main content of acceptance self-inspection report are as follows:

1 The preparation requirements for and the main content of design self-inspection report are shown in Appendix F of this specification.

2 The main content of supervision self-inspection report is shown in Appendix G of this specification.

3 The main content of construction and installation self-inspection report is shown in Appendix H of this specification.

4 The main content of operation self-inspection report is shown in Appendix J of this specification.

6.0.3 The content and level of detail of safety acceptance assessment report shall comply with the current sector standard NB/T 31027, *Code for Preparation of Safety Assessment Report upon Completion of Wind Power Projects.*

6.0.4 The project owner organizes the preparation of acceptance documents, which shall be submitted by the relevant organizations on time as required. The project owner shall check the completeness and standardization of the submitted acceptance documents.

6.0.5 Acceptance documents include technical documents and documents for reference. The relevant organizations shall ensure the authenticity of the documents submitted and bear the corresponding accountability.

6.0.6 The drawings, data and deliverable documents used for acceptance shall be prepared as required. Except for drawings, acceptance documents shall be prepared in the international standard format of A4 (210 mm × 297 mm). The original text shall be stamped with the official seal, and no photocopies shall be adopted.

Appendix A Occupational Safety and Health Acceptance Procudure for Wind Power Projects

A.0.1 The occupational safety and health acceptance procedure for wind power projects is shown in Figure A.0.1.

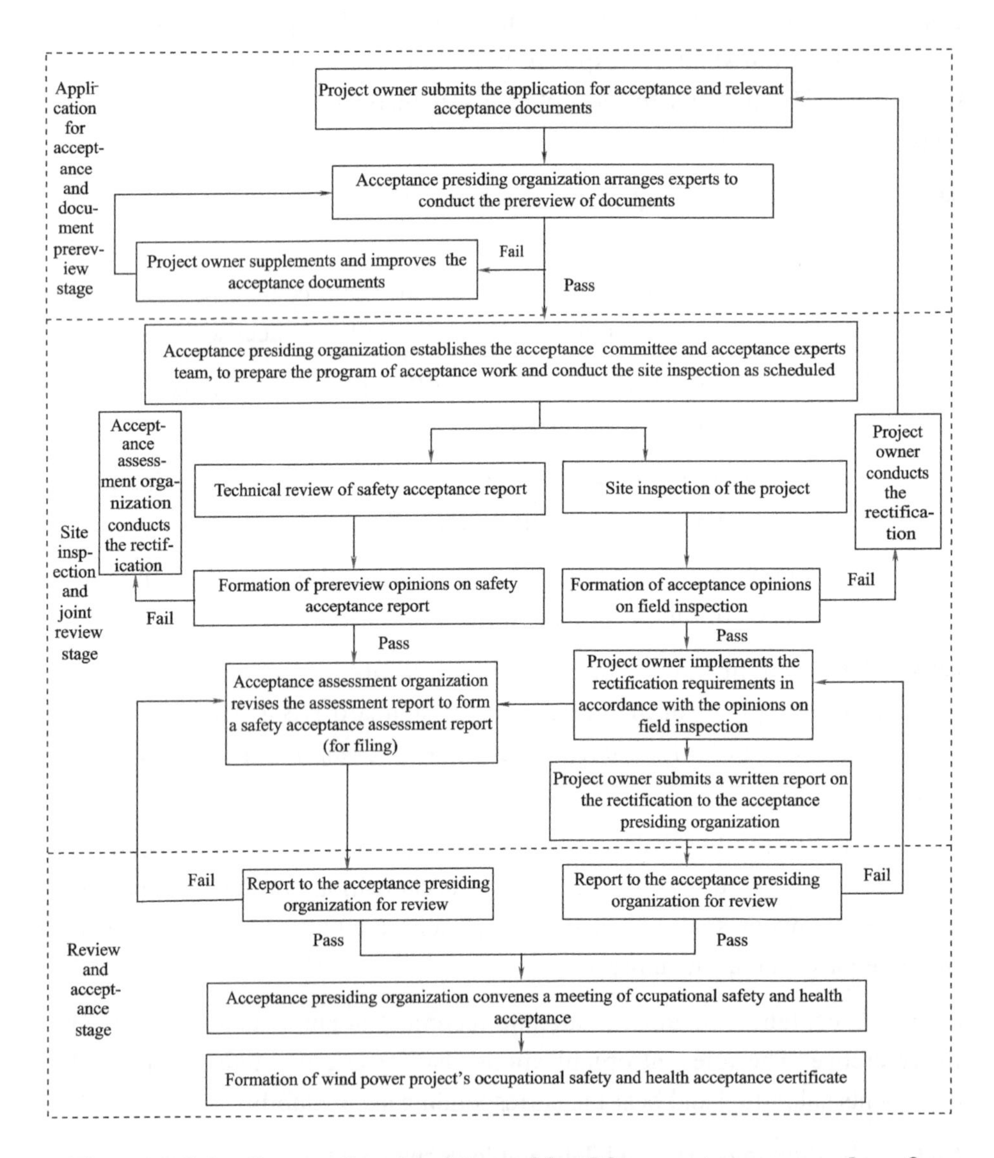

Figure A.0.1 Occupational safety and health acceptance procedure for wind power projects

Appendix B Application Form for Occupational Safety and Health Acceptance of Wind Power Projects (Format)

B.0.1 The application form for occupational safety and health acceptance of wind power projects is shown in Table B.0.1.

Table B.0.1 Application form for occupational safety and health acceptance of ×× project

<table>
<tr><td colspan="6">Ⅰ. Project Owner Information</td></tr>
<tr><td>Name</td><td colspan="3"></td><td>Nature</td><td></td></tr>
<tr><td>Address</td><td colspan="3"></td><td>Postcode</td><td></td></tr>
<tr><td>Legal Representative</td><td></td><td>Tel.</td><td></td><td>Fax</td><td></td></tr>
<tr><td rowspan="2">Contact Person</td><td rowspan="2"></td><td>Tel.</td><td></td><td>Fax</td><td></td></tr>
<tr><td>E-mail</td><td colspan="3"></td></tr>
<tr><td colspan="6">Profile:</td></tr>
<tr><td colspan="6">Ⅱ. Project Information</td></tr>
<tr><td>Name</td><td colspan="5"></td></tr>
<tr><td>Investment (ten thousand yuan)</td><td colspan="2"></td><td>Installed Capacity (kW)</td><td colspan="2"></td></tr>
<tr><td colspan="4">Staffing (person)</td><td colspan="2"></td></tr>
<tr><td colspan="4">Commissioning date (MM/DD/YYYY)</td><td colspan="2"></td></tr>
<tr><td colspan="4">Expected date of putting into operation (MM/DD/YYYY)</td><td colspan="2"></td></tr>
<tr><td colspan="6">Profile:</td></tr>
</table>

Table B.0.1 *(continued)*

Ⅲ. Design information				
Designer	Name			
	Correspondence Address			
	Contact			
	Qualification Certificate Number			
Legal Representative	Name		Tel.	
	Fax		E-mail	
Design Completion Date	MM/DD/YYYY			
Profile:				

Table B.0.1 *(continued)*

<table>
<tr><td colspan="6">Ⅳ. Project Safety Acceptance Assessment</td></tr>
<tr><td rowspan="3">Assessment Organization</td><td>Name</td><td colspan="4"></td></tr>
<tr><td>Correspondence Address</td><td colspan="4"></td></tr>
<tr><td>Qualification Certificate Number</td><td colspan="4"></td></tr>
<tr><td rowspan="2">Legal Representative of Assessment Organization</td><td>Name</td><td colspan="2"></td><td>Tel.</td><td></td></tr>
<tr><td>Fax</td><td colspan="2"></td><td>E-mail</td><td></td></tr>
<tr><td rowspan="2">Contact Person</td><td rowspan="2"></td><td>Tel.</td><td></td><td>Fax</td><td></td></tr>
<tr><td>E-mail</td><td colspan="3"></td></tr>
<tr><td rowspan="4">Safety Acceptance Assessment Report</td><td>Person in Charge of Assessment</td><td colspan="2"></td><td>Tel.</td><td></td></tr>
<tr><td>Report Name</td><td colspan="4"></td></tr>
<tr><td>Report Completion Date</td><td colspan="4">MM/DD/YYYY</td></tr>
<tr><td colspan="5">Conclusion:</td></tr>
<tr><td colspan="3">Safety Acceptance Assessment Organization:

(Seal)
MM/DD/YYYY</td><td colspan="3">Project Owner:

(Seal)
MM/DD/YYYY</td></tr>
</table>

Appendix C Main Content of Occupational Safety and Health Acceptance Program

C.0.1 Project overview.

The geographical location, project scale, construction schedule, owner information, construction participants, implementation of occupational safety and health acceptance-related work, main approval documents obtained, etc.

C.0.2 Main basis for acceptance work:

1 Main laws, regulations, and regulatory documents applicable to the acceptance work.

2 Technical documents to be reviewed for acceptance.

3 Relevant approval documents obtained.

C.0.3 Acceptance organization:

1 Acceptance leading organization. Establish the acceptance committee, and clearly define the membership.

2 Acceptance working organization and responsibilities. Clearly define the acceptance working organization and its responsibilities.

3 Participating organizations in acceptance. Clearly define the acceptance participants, mainly including the local work safety supervision authority, acceptance committee member organizations, construction participants, and safety assessment organization.

C.0.4 Acceptance conditions and basic documents:

1 Basic condition required for the project acceptance.

2 Basic documents that shall be prepared by project owner. Clearly define the acceptance basis report, and the basic documents that shall be prepared by project owner (including information for reference).

C.0.5 Acceptance principles, acceptance scope, and main work content.

C.0.6 Acceptance procedures, and scheduling.

C.0.7 Principles and methods to address the issues arising from the acceptance work.

Appendix D Appraisal Certificate of Occupational Safety and Health Acceptance for Wind Power Projects

D.0.1 The front cover of the appraisal certificate is shown in Figure D.0.1:

××Project

Occupational Safety and Health Acceptance

(18-point FangSong GB 2312)

Appraisal Certificate

(36-point FangSong GB 2312)

Acceptance Committee for Occupational Safety and Health Date

(15-point FangSong GB 2312)

Figure D.0.1 Front cover of the appraisal certificate

D.0.2 The appraisal certificate of occupational safety and health acceptance for wind power projects consists of the following:

1 Foreword. Acceptance basis, acceptance organizer and participants, and briefing of acceptance meeting.

2 Project overview:

1) Project name, geographical location, names of main buildings, project scale, main indicators of comprehensive economic benefits (design parameters), etc.

2) Briefing of investigation and design, including feasibility study report review or approval, project approval, etc.

3) Full names of the relevant organizations including designer, construction contractor, supervisor, equipment manufacturer, and operator, and their main tasks.

3 Acceptance principles and acceptance scope.

4 Main dangerous and hazardous factors and their criticality.

5 Implementation of “regime of three concurrences”.

6 Major design changes relating to safety.

7 Main safety facilities.

8 Safety management.

9 Conclusion of project safety acceptance assessment report.

10 Problems in existence and recommended solutions.

11 Acceptance conclusions.

12 Attachments:

Attachment 1: Signature Form of Acceptance Committee Members for Occupational Safety and Health Acceptance of ×× Project (discrepancy shall be signed by the dissenter).

Attachment 2: Signature Form of Experts Team Members for Occupational Safety and Health Acceptance of ×× Project.

Attachment 3: Signature Form of the Representatives from the Construction Participants for Occupational Safety and Health Acceptance of ×× Project.

Attachment 4: List of Documents and Information for Occupational Safety and Health Acceptance of ×× Project.

Appendix E Preparation Requirements for Self-Inspection Report of Occupational Safety and Health Acceptance

E.0.1 The self-inspection report for acceptance consists of the following:

1 Front cover (see Figure E.0.1-1).

2 Inner cover (see Figure E.0.1-2).

3 Contents.

4 Main body.

5 Attachments.

E.0.2 The font size and font type are as follows:

1 Headings in the main body:

1) Chapter and section headings are in 16-point SimHei and KaiTi font respectively.

2) Item headings are in 14-point SimHei font.

2 Content in the main body:

1) The text is in 14-point SimSun font.

2) The words in tables may be in either 10.5-point or 7.5-point SimSun font.

E.0.3 Paper and layout requirements are as follows:

1 A4 white offset paper (70 g or above).

2 The left margin is 28 mm, the right margin is 20 mm, the top margin is 25 mm, and the bottom margin is 20 mm.

3 The chapter and section headings are centered, and two blank spaces are indented for item headings.

E.0.4 Printing requirements are as follows:

Printed on both sides except for the attached drawings, photocopies, etc.

E.0.5 The cover page requirements are as follows:

The front cover of the acceptance self-inspection report is stamped with the official seal of the preparation organization, and the input item shall be signed by the responsible person.

× × Project
Occupational Safety and Health Acceptance
(18-point FangSong GB 2312)

Self-Inspection Report

(36-point FangSong GB 2312)

(Design, Supervision, Construction, and Operation)
(16-point FangSong GB 2312)

Name of the preparation organization (official seal)
Date
(15-point FangSong GB 2312)

Figure E.0.1-1 Front cover of the self-inspection report

× ×Project
Occupational Safety and Health
Acceptance
(18-point FangSong GB 2312)

(Design, Supervision, Construction, and Operation)
Self-Inspection Report
(24-point FangSong GB 2312)

Approved by:
Checked by:
Reviewed by:
Prepared by:
(16-point FangSong GB 2312)

Figure E.0.1-2 Inner cover of the self-inspection report

Appendix F Preparation Requirements and Main Content of Design Self-Inspection Report

F.0.1 The basic preparation requirements for the design self-inspection report are as follows:

1 Describe the safety technologies and facilities that have been adopted in the project design for the sake of ensuring the long-term safe operation of the project after the completion; summarize the lessons learned and experience gained from the implementation of the project.

2 Evaluate the safety condition of the project from the perspective of engineering design, point out the safety hazards, and propose the recommended countermeasures.

F.0.2 Design self-inspection report shall include the following:

1 Design basis:

1) Applicable regulations of the national, industry-specific, local and project authorities.

2) Main technical standards adopted in the design.

3) Design scope of engineering protection.

2 Safety design for the prevention of major dangerous and hazardous factors and its assessment:

1) Safety design for the prevention of major dangerous and hazardous factors in the wind farm site selection and general layout and its assessment.

2) Safety design for the prevention of dangerous and hazardous factors in wind turbines, collection lines, step-up sub-stations, etc. and its assessment.

3) Safety design for the prevention of major dangerous and hazardous factors in the operation process, and its assessment.

Including the engineering safety measures cover the fire protection and explosion prevention; the prevention of equipment defects, protection defects, signal defects, sign defects, electrical hazards (exposure of live parts, electric leakage, lightning, static electricity, electric sparks, etc.); moving object hazards; the injuries caused by falls from elevated position, object hitting, lifting, vehicle and machinery; the prevention of floods, drowning, collapse, intoxication and asphyxiation, etc. in the production process for production equipment and installations, and main structures.

4) Safety design of special-purpose equipment and its assessment.

5) Safety design for the prevention of environmental hazardous factors in the workplace within the wind farm area, and its assessment.

Including the safety measures for the prevention of noise, vibration, dust and toxics; and heating, ventilation and air conditioning (HVAC), daylighting, lighting, etc.

6) Safety design for the prevention of major dangerous and hazardous factors during the construction period.

3 Design of safety signs and safety markings.

4 Safety monitoring system and main instruments and equipment configuration.

5 Major safety-related design changes and safety measures.

6 Main conclusions and recommendations.

7 Attached drawings:

1) Geographic location map of the wind farm.

2) General layout plan of the wind power project.

3) General layout plan of the step-up substation and main structures (buildings).

4) Layout of wind turbines of the wind farm.

5) Foundation design drawing of wind turbine and box-type transformer.

6) Main electrical connection of the step-up sub-station.

7) Collection line drawing of the wind farm.

8) Layout of the ultra-high voltage electrical equipment in the step-up substation.

9) Layout of the electrical equipment on each floor of the main control building.

10) General layout plan of construction.

Appendix G Main Content of Supervision Self-Inspection Report

G.0.1 Brief description of the quality and safety assurance system of the supervision organization and its implementation.

G.0.2 The basic information on the quality supervision and control (including the quality of permanent equipment) in the process of civil construction and equipment installation of the project.

G.0.3 The acceptance ratings of works (sub-works) and concealed works and the rectification and treatment of outstanding problems.

G.0.4 The handling of major quality and safety accidents and the implementation of major design changes during the construction period, and the rectification effect assessment on the possible remaining parts with quality and safety hazards.

G.0.5 The review of the technical safety measures or special construction schemes in the construction planning and the solutions of problems in existence. The compliance of the technical safety measures or special construction schemes with the mandatory standards for construction.

G.0.6 Description on the construction and implementation of safety technologies and safety facilities proposed in the project design, as well as the possible outstanding safety problems.

G.0.7 Description on the handover acceptance inspection of safety technologies and safety facilities organized by the supervision organization, especially the acceptance inspection data of the works (sub-works) that is required to be tested for a full commissioning period.

G.0.8 Conclusion on whether the project safety facilities have been constructed according to the design documents, and the construction quality meets the requirements specified in the design documents of the project safety facilities.

G.0.9 Conclusion on whether the construction of project safety facilities meets the requirements of the relevant construction standards of China.

Appendix H Main Content of Construction and Installation Self-Inspection Report

H.0.1 Briefly describe the quality and safety assurance system of the civil construction contractor and equipment installation contractor and its implementation.

H.0.2 Describe the quality control in the civil construction and equipment installation process (including the quality of permanent equipment), the quality assurance measures of works (sub-works), the rectification and treatment of quality problems related to the installation of civil construction facilities and electromechanical equipment, the handling of major quality and safety accidents and implementation of major design changes during construction, especially the problems in existence and their impact on the project.

H.0.3 Describe the quality problems of safety technologies and safety facilities proposed in engineering design occurred in the construction, installation, commissioning and trial operation, as well as problem analysis, implementation and treatment process.

H.0.4 Describe the major hazards in the safety facilities found during construction and installation and their handling measures.

H.0.5 Describe the handover acceptance inspection of the works relating to safety technologies and safety facilities, and conduct a comprehensive evaluation on the construction and installation quality of civil works and equipment installation works (including the excellent rate, qualification rate, etc.)

Appendix J Main Content of Operation Self-Inspection Report

J.0.1 Project overview and commissioning.

Briefly describe the wind power project construction and the wind turbines commissioning, preparation process, staffing, education and training, the participation in the entire process of supervision, construction, installation, commissioning and trial operation; elaborate the structures (buildings), equipment and power grid accidents happened since the trial operation and their consequences (casualties and loss of equipment and property), and the technical measures against accident.

J.0.2 Inspect the implementation of "regime of three concurrences".

Inspect and describe the implementation of the national and local requirements for the "regime of three concurrences" related to safety facilities in the wind power project construction. Describe the implementation of main countermeasures and recommendations proposed in the safety pre-assessment report (for filing) and review comments, chapter of occupational safety and health design in the feasibility study report, and the safety acceptance assessment report of the wind power project.

J.0.3 The rectification of the problems found in other special-item acceptance.

Present the problems found in the special-item acceptance such as fire protection of the project and their rectification.

J.0.4 The safety inspection of trial operation status.

According to the actual experience in the trial operation management since the wind turbines were gradually put into operation, conduct the comprehensive inspection on the project site selection and general layout, measures against accidents for operation structures (buildings) and equipment, prevention of dangerous factors during production and operation, working environment (hazardous factors or occupational health in workplaces) and conditions in the wind farm area, engineering safety monitoring system, special-purpose equipment safety and work safety management, etc.; point out the hazards that might endanger personnel and equipment safety; propose the safety prevention and measures against accidents that shall be taken before the completion acceptance of the project; mainly including:

1 The implementation of safety measures and the configuration and operation of the safety monitoring system for the prevention of geological hazards, building collapse, equipment fault, fire,

explosion, electrical injury, lifting injury, maloperation, fall accidents, vehicle injury, mechanical injury, overhead line fault, equipment commissioning, natural disasters, offshore wind turbine corrosion, scouring, equipment maintenance, etc.; provide a list of main safety facilities and equipment, operational monitoring records and results analysis report.

2 The actual operation situation of facilities and equipment for the prevention of high temperature, low temperature, noise, electromagnetic field, hazardous substances, salt spray corrosion, sand blown by wind, typhoons, etc.; provide a list of main equipment and the detection report of hazardous factors in workplaces.

3 Whether the on-site access, emergency rescue, safety signs, ventilation and other facilities in the workplaces comply with national and industry regulations.

4 The inspection of lightning protection system, earthing resistance of wind turbines, foundation settlement of each wind turbine, heating and cooling devices of wind turbine, safety control system of wind turbines, communication channel of remote control system, electrical insulation tools and safety tools for climbing; provide the inspection records and values; the inspection of safety wire rope, climbing ladder, working platform and door windproof hook inside the tower, and their inspection records.

5 Conduct the self-inspection on the trial operation of the project against the safety checklist.

J.0.5 Ancillary safety and health facilities.

Describe the configuration and use of the main ancillary safety and health facilities of the project.

J.0.6 Work safety management.

Describe the implementation of work safety management organization, safety production responsibility system at all levels, operation ticket management, work ticket management, safety management of special operation personnel, safety management of work at height, equipment management, safety apparatus management, PPE management, management of measures against accidents and technical safety measures, emergency management, employee work safety education and training, management of rewards and punishments for safe and civilized production, regular safety meetings, safety inspection management, safety management of contracted works, safety management of temporary workers, work safety archives management, accident investigation management, fire protection management, maintenance and repair management, etc.; provide the lists of work safety management systems.

J.0.7 Management of major hazards.

State the distribution of major hazard sources, the testing, inspection, and maintenance of safety facilities and safety monitoring systems with major hazards, the emergency management of major hazards, and the registration and filing of major hazards according to the current national standards, regulations and normative documents of government authorities and relevant requirements of the competent administrative department for enterprises.

J.0.8 Preparation and implementation of emergency preparedness plan.

Describe the emergency preparedness plan of the project, the connection with related emergency preparedness plans, emergency drills, and implementation in accordance with the current national standard GB/T 29639, *Guidelines for Enterprises to Develop Emergency Response Plan for Work Place Accidents.*

J.0.9 The implementation of the cost estimation for occupational safety and health.

Describe the implementation of the special investment proposed in the safety pre-assessment report and the chapter of "Occupational safety and health" design of the project. Including the technical safety measures, PPE issued to workers according to national standards, the work safety education and training for workers, the configuration of safety monitoring equipment, devices, facilities, instruments and meters, the preparation of emergency preparedness plan and its drill, materials in place, and the use of relevant funds.

Explanation of Wording in This Specification

1 Words used for different degrees of strictness are explained as follows in order to mark the differences in executing the requirements in this specification.

1) Words denoting a very strict or mandatory requirement:

"Must" is used for affirmation; "must not" for negation.

2) Words denoting a strict requirement under normal conditions:

"Shall" is used for affirmation; "shall not" for negation.

3) Words denoting a permission of a slight choice or an indication of the most suitable choice when conditions permit:

"Should" is used for affirmation; "should not" for negation.

4) "May" is used to express the option available, sometimes with the conditional permit.

2 "Shall meet the requirements of..." or "shall comply with..." is used in this specification to indicate that it is necessary to comply with the requirements stipulated in other relative standards and codes.

List of Quoted Standards

GB/T 29639, *Guidelines for Enterprises to Develop Emergency Response Plan for Work Place Accidents*

NB/T 31027, *Code for Preparation of Safety Assessment Report upon Completion of Wind Power Projects*